Columba Sara Evelyn (Ed.)

LBJ/Central (DART Station)

Columba Sara Evelyn (Ed.)

LBJ/Central (DART Station)

DART Light Rail, High Five Interchange, Interstate 635 (Texas), Central Expressway (Dallas)

Fec Publishing

Publisher:
Fec Publishing is a trademark of
International Book Market Service Ltd., 17 Rue Meldrum, Beau Bassin, 1713-01 Mauritius
Email: info@bookmarketservice.com
Website: www.bookmarketservice.com

Published in 2011

Printed in: U.S.A., U.K., Germany. This book was not produced in Mauritius.

ISBN: 978-613-9-99994-1

Contents

References

LBJ/Central_(DART_station)

LBJ/Central

DART light rail station

Station statistics	
Address	8902 Lyndon B. Johnson Freeway, Dallas, Texas 75243
Coordinates	32°55′6″N 96°45′7″W
Lines	Red Line Orange Line
Connections	DART Routes 234 [1] (M-F), 567 [2], and 582 [3] DART TI North Shuttle [4] (M-F) DART TI South Shuttle [5] (M-F) DART TI Forest Lane Shuttle [6] (M-F)
Structure	At grade
Platforms	Two side platforms
Parking	568 free spaces, no overnight
Bicycle facilities	One locker, one rack
Other information	
Opened	1 July 2002
Accessible	♿
Owned by	Dallas Area Rapid Transit

Services

Preceding station	Dallas Area Rapid Transit	Following station
Forest Lane *toward Westmoreland*	Red Line	Spring Valley *toward Parker Road*
Forest Lane *toward Bachman*	Orange Line	
	Effective Late 2011	
Forest Lane *toward Irving Convention Center*	Orange Line Parker Road branch	Spring Valley *toward Parker Road*

LBJ/Central Station is a DART Light Rail station located in the Lake Highlands area of Dallas, Texas (USA) at T.I. Boulevard and Markville Drive near the High Five Interchange where Interstate 635 (LBJ Freeway) and

US 75 (North Central Expressway) meet. It opened 1 July 2002 and is a station on the Red Line, serving the headquarters of Texas Instruments, offices of Raytheon Company, and nearby Richland College.

External links

- DART - LBJ/Central Station [7]

DART_Light_Rail

Dallas Area Rapid Transit Light Rail	

DART train stopped at the West End Station, heading to the American Airlines Center.

Info	
Owner	DART
Locale	Dallas, Texas
Transit type	Light rail
Number of stations	54 (open) 62 (open + under construction)
Operation	
Began operation	June 14, 1996
Technical	
Track gauge	1435 mm (4 ft $8\frac{1}{2}$ in)
Minimum radius of curvature	(?)
Electrification	750 V DC, overhead lines

DART Light Rail is a light rail system in Dallas, Texas and its suburbs owned by the Dallas Area Rapid Transit. The system comprises 72 miles (115.9 km) between its three lines — the Red Line, the Blue Line and the Green Line. Its 72 miles of track make it the largest light rail system in the United States[1] . It has a daily ridership of 75,000 average trips per weekday[2] . The system utilizes light rail trains manufactured by Kinki Sharyo, with all trains being converted to "Super" LRVs which feature level boarding (especially convenient for strollers and bicycles and increasing the capacity for wheelchairs from about 4 per train to 6-8 per car) and higher passenger capacity.[3]

History

Before the 1983 election, DART had a plan for 160 miles (257.5 km) of rail. After the election, the plan was pared down to 147 miles (236.6 km) when Duncanville, Grand Prairie and Mesquite, which would have had rail lines, didn't opt to join the agency. DART chose light rail transit as its primary mode of rail transportation in 1984. The plan was pared down again to 93 miles (149.7 km) before the 1988 bond vote. After the vote, the agency again pared down the regional rail system to 84 miles (135.2 km); 66 miles (106.2 km) light rail and 18 miles of commuter rail. In 2010, DART has announced that it will likely need to scrap much of its 2030 plan for expansion citing deficits and money woes.[4]

The following lines are currently active:

- Red Line (Opened in 1996, completed in its current state in 2002)
- Blue Line (Opened in 1996, completed in its current state in 2002, additions scheduled for completion in 2012 and 2018)
- Green Line (Opened in 2009, completed in its current state in 2010)

The following line is under construction:

- Orange Line (Irving/DFW Line; proposed opening 2011, DART may need to scrap the extension from Irving to DFW Airport to save $50 million. [5])

Starter system

After six years of scandals involving finances, delays in construction of the rail system, a failed bond election, and nine pull-out votes (two of which were successful), the year 1990 was a turning point for the agency when DART broke ground on its light rail system in October. The light rail system began service on time and on budget in June 1996,[6] inaugurating the first light rail system in Texas and the Southwest. Commuter rail service to Irving began in December, after some delays.

Train arriving at the Downtown Plano Station

To the surprise of critics, who expected the low ridership of some other systems, the new light rail system was embraced by Dallasites, with ridership exceeding expectations. The suburbs' confidence in DART was also expressed at the ballot box: four cities held highly publicized pullout elections in 1996 (with the financial assistance of Dallas Cowboys owner Jerry Jones who wanted DART's half-cent sales tax revenue for Cowboy Stadium then in Irving [7]), but all four voted to remain in DART (three of them by margins of more than two-to-one).

The 20 miles (32 km) starter system opened on 14 June 1996, which spanned from the Westmoreland Station on the Red Line in Dallas (West Oak Cliff) through downtown Dallas to Pearl Station. There were also two stations on the Blue Line, Morrell and Illinois that opened on that date.

The next stage of the starter line opened on 10 January 1997 when the Red Line was extended from Pearl Station to Park Lane Station. It includes the underground portion of the line between Pearl Station and Mockingbird Station which the site of Dallas's first modern transit village. The northern terminus of the Blue Line was still Pearl Station.[8]

The starter line was completed when the Blue Line was extended from Illinois Station to its current southern terminus at Ledbetter Station. This came on 31 May 1997, nearly one year after the first part of the starter system opened.

The final change to the starter system came on 18 December 2000 when DART opened Cityplace Station, the first subway station in the Southwest, located in the tunnel between Pearl and Mockingbird station under the Tower at

Cityplace skyscraper, providing access to the Tower as well as the never built Cityplace West.

Commuter Rail on the old Rock Island right-of-way began during the series of openings of the light rail starter system. On 30 December 1996 the Trinity Railway Express opened the first segment as the South Irving Transit Center was connected to Union Station in downtown Dallas with a stop in the middle at Medical/Market Center Station. The agency had to lease rail cars due to a delay in the current stock until March.

Suburban expansion

After the success of the light rail starter system, voters approved DART's request to use long-term bonds to expedite the construction of the regional light rail system. While DART had originally planned single-track extensions of the Blue Line to Garland and the Red Line to Richardson and Plano, the starter system was so popular that the agency made plans to double-track the entire route.

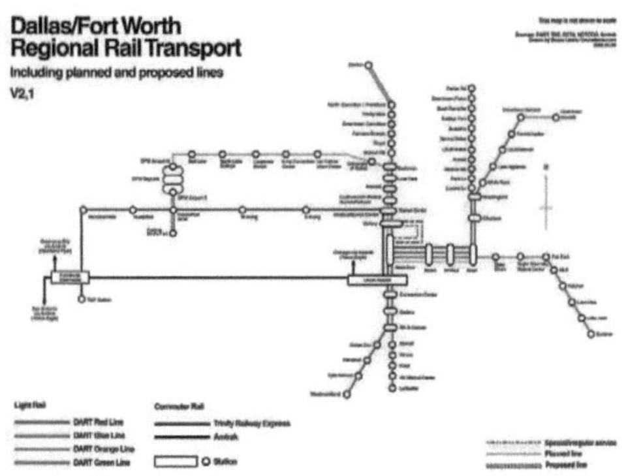

DFW map of proposed system

Work began on the extensions of the Blue Line and the Red Line on 15 January 1999. Later that same year, on 27 September, the Blue Line designation was extended from Pearl Station to Mockingbird Station, for its eventual expansion to Garland.[8] The first station since 1997 to open came on 24 September 2001, when the Blue Line was extended from Mockingbird Station to the new White Rock Station. A little over a year and a half later, the Blue Line debuted another station as light rail pushed outside the LBJ Loop when the LBJ/Skillman Station opened on 6 May 2002.

The first extension of the Red Line opened on 1 July 2002 when the line added 7 stations from a newly opened Park Lane Station to Galatyn Park Station. Richardson became the first Dallas suburb to be served by light rail.

The Blue Line was completed in its current form on 18 November 2002 when Garland became the second suburb to get light rail service. The two new stations of Forest/Jupiter Station and Downtown Garland Station were opened to the public.

The final stage of the light rail expansion was completed six months ahead of schedule when Plano's first trains ran on 9 December 2002, completing the north central and northeast expansion. Archer Western Contractors is the Construction Management and General Contractor officials building the current rail extensions to Carrollton and Elam Creek.

Passengers waiting at the Hampton Station.

Green and Orange Lines

On 3 July 2006, the Federal Transit Administration (FTA) approved a US$700 million Full Funding Grant Agreement (FFGA) — the largest grant ever awarded to DART — to kick-start a US$2.5 billion expansion of the light rail system. Now under construction, this phase will include two new light rail lines that will double DART's existing light rail mileage.[9] Construction began in September 2006. When the project is complete in 2013, DART's light rail system will have doubled to 90 miles (145 km).[10]

The Green Line began operation on September 14, 2009 with a route from downtown Dallas southeast to Fair Park; this short route was scheduled to open in time to service the 2009 State Fair of Texas. On December 6, 2010 the line extended further to Pleasant Grove, as well as continuing northwest from Victory Station to Farmers Branch and Carrollton; both extensions, completing the Green Line.[11]

The northern end of the line connects with the A-train line run by the Denton County Transportation Authority (DCTA).[12] This line connects Denton to Carrollton, with stops in Lewisville and Highland Village. The commuter train may stop in other Denton County cities, should they choose to join the DCTA.

The second line, the Orange Line will run concurrently with the Green Line from downtown Dallas to just past Love Field Airport, where it will branch off toward Irving with eventual service to DFW Airport. DART is cooperating with DFW and Love Field to link the two airports via rail. The line will run through Las Colinas and connect to the Las Colinas APT System. DART had preliminary plans for the Orange Line to run concurrently with the Red Line from downtown Dallas to LBJ/Central Station.[13] Current maps have the Orange Line's terminus at Pearl Station in downtown Dallas.

In maps before 2006, DART labeled the Pleasant Grove to Carrollton route the "Orange Line", and the Irving route was the "Purple Line". Green was generally used on DART maps to denote the route of the Trinity Railway Express. By the time construction started, DART was using the new "Green Line" designation as part of its marketing efforts, saying "Like the color green, this line is a symbol of our city on the move."[14]

2030 Plan

In 2010, citing deficits and drops in revenue, DART has scrapped much of their 2030 plan.[15]

In October 2006, the DART Board of Directors unanimously approved the following measures for the next round of rail expansion.[16]

- A 2.9-mile (4.7 km) light rail extension of the Blue Line to Bonnie View Road and Interstate 20 to the new SouthPort intermodal port in southeast Dallas. This line will terminate just north of Lancaster and Hutchins, cities that are not currently members of DART.
- A 4.3-mile (6.9 km) light rail extension of the Red Line south to Red Bird Lane. This line will terminate just short of Duncanville, another non-DART city, and pass near the headquarters of AAFES.

- A 4.3-mile (6.9 km) light rail extension of the Orange Line along Scyene Road to approximately Masters Drive, just short of non-DART member city Mesquite.
- A 6-mile (10 km) light rail line in West Dallas along Fort Worth Avenue or Singleton to Loop 12/Jefferson Boulevard, terminating just east of Grand Prairie, which also did not join DART. No color designation has been given for this planned line.
- A station for the Lake Highlands neighborhood of northeast Dallas on the Blue Line, between White Rock Station and LBJ/Skillman Station. The area previously opposed rail service, so the initial line was built through the area without a station at the turn of the century. The new station opened on December 6, 2010 becoming the first component of the 2030 plan to be completed and the first infill station in DART's system.[17] This is a key component of the new Lake Highlands Town Center TOD development.[18]
- A nearly 26-mile (42 km) Cotton Belt Rail Line commuter service in the east-west Cotton Belt corridor from the Red Line in Plano to DFW International Airport. This line would provide rail service to the existing bus transfer station in Addison and would intersect the Green Line at Downtown Carrollton Station. The line would pass through Coppell, which pulled out of DART in 1989. No color designation has been given for this planned line.

DART's 2030 plan also calls for a comprehensive network of enhanced and rapid bus corridors consisting of 77 miles (124 km) of enhanced bus service corridors and 20 miles (32 km) of rapid bus service corridors with strengthened and new express bus service. The plan also calls for additions to the current 31 miles (50 km) of high occupancy vehicle lanes to 116 miles (187 km).

The final 2030 plan included several changes from the draft plan released in July 2006. Removed from the final plan was a 6.3-mile (10.1 km) branch of the Red Line from Forest Lane Station to the Addison Transit Center, which would have included several miles of subway under Interstate 635. DART officials cited the line's high cost, US$700 million, and lack of strong support from the city of Dallas. The Blue Line extension to the Dallas Southport Center (intermodal terminal) was added after strong pressure from Dallas officials. Also, the proposed light rail line serving West Dallas was not originally considered as a priority for rail service.

The Cotton Belt corridor plans continued to generate controversy right up to the day of the vote on the 2030 plan. DART leaned toward diesel powered commuter rail for the Cotton Belt corridor, similar to the Trinity Railway Express. However, the line would pass through affluent Far North Dallas neighborhoods. The neighborhood formed the Cotton Belt Concerned Coalition to fight the line in 1990. The group made a concession in 2006 to allow electric light rail vehicles on the line to avoid perceived air and noise pollution associated with diesel rail. The group also proposed that the line be placed in a below-ground trench. These proposals were accepted by the City of Dallas in June 2006 in a unanimous resolution. DART, however, balked at the cost of trenching the line, which they estimated at $250 million. This set up a confrontation between DART and the city of Dallas, which appoints eight of the board's 15 members. The final plan compromised by promising $50 million "to help address neighborhood concerns".[19]

The current freight line runs just south of the Downtown Plano Station on the Red Line, but DART maps of the 2030 plan show the line will route to the nearby Bush Turnpike station, the nearest station to the south of downtown Plano.[20] The Cotton Belt line runs through former DART member city Coppell between Carrollton and DFW Airport; although no station locations are included in the plan, the promise of a future station could entice Coppell into rejoining the agency.

See also

- List of DART Light Rail stations
- List of tram and light-rail transit systems
- List of United States light rail systems by ridership
- McKinney Avenue Transit Authority

References

[1] http://www.houstontomorrow.org/livability/story/dallas-green-line-begins-service/
[2] http://www.dart.org/news/news.asp?id=963
[3] Facts:SLRV (DART Fact Sheets) (http://www.dart.org/factsheet/SLRV)
[4] Money woes will force DART to cut jobs, rail plans (http://www.dallasnews.com/sharedcontent/dws/dn/latestnews/stories/042710dnmetdart.37eb390.html) Dallas Morning News
[5] http://www.dallasnews.com/sharedcontent/dws/dn/latestnews/stories/042710dnmetdart.37eb390.html
[6] Kelley, Chris (June 16, 1996). "Last original staffer delivered train on time, within budget". *The Dallas Morning News*, p. 31A.
[7] http://www.bizjournals.com/dallas/stories/1996/07/22/editorial2.html Keep DART in Irving - Dallas Business Journal
[8] Dallas Light Rail (http://web.presby.edu/~jtbell/transit/Dallas/DART/)
[9] http://www.dart.org/news.asp?ID=714
[10] http://www.dart.org/mis.asp?zeon=DARTRailDouble
[11] Solis, Dianne (June 10, 2009). "Officials reshaping downtown Carrollton around light-rail hopes" (http://www.dallasnews.com/sharedcontent/dws/news/localnews/stories/060909dnmetoldtown.4a65d43.html). *Dallas Morning News*. . Retrieved June 10, 2009.
[12] Peterson, Matt (June 20, 2011). "A-train railway begins rolling, carrying commuters from Denton to Carrollton" (http://thescoopblog.dallasnews.com/archives/2011/06/a-train-railway-begins-rolling.html). *The Dallas Morning News*. . Retrieved June 20, 2011.
[13] NW Corridor FEIS, Figure 2-19 LRT Operating Plan (http://www.dart.org/nweis/nwfeisfigures.htm)
[14] Community Meeting/Public Hearing Notices - DART Green Line Expansion; 16 November 2006 (http://www.dart.org/publicmeetings.asp?ID=267)
[15] "Money woes will force DART to cut jobs, rail plans," by Michael A. Lindberger (The Dallas Morning News; Wednesday, April 28, 2010) (http://www.dallasnews.com/sharedcontent/dws/dn/latestnews/stories/042710dnmetdart.37eb390.html)
[16] DART's 2030 plan (http://www.dart.org/news.asp?ID=719)
[17] DART.org - Blue Line Lake Highlands Station (http://www.dart.org/about/expansion/bluelinelakehighlandsstation.asp) (accessed November 27, 2010)
[18] (http://www.lakehighlandstowncenter.com/synopsis.html)
[19] DART green-lights 40 miles of new rail for 2030 (http://www.dallasnews.com/sharedcontent/dws/news/localnews/stories/DN-dart_25met.ART0.North.Edition1.3e3a81a.html)
[20] DART 2030 plan map (http://www.dart.org/2030plan/DART2030PlanJan2007.pdf)

External links

- Official site (http://www.dart.org/)
- Official site (Spanish Language) (http://www.transportedart.org)
  - System Map (http://www.dart.org/maps/systemmap.asp)
  - Expansion Plan Map (http://www.dart.org/maps/currentandfutureservicesmap.asp)
  - DART History (http://www.dart.org/about/history.asp)

High_Five_Interchange

High Five Interchange	
 The High Five under construction.	
Location	
Dallas, Texas	
Coordinates:	32°55′27″N 96°45′49″W
Roads at junction:	US 75 (Central Expressway) I-635 (LBJ Freeway)
Construction	
Type:	Five-level stack interchange
Constructed:	2002–2005
Opened:	December 2005
Maximum height:	120 ft (37 m)
Maintained by:	the Texas Department of Transportation

The **High Five Interchange** is a five-level freeway interchange, designed by the HNTB Corporation,[1] that connects Lyndon B. Johnson (LBJ) Freeway (Interstate 635) and Central Expressway (US 75) in Dallas, Texas, United States. The High Five is the first five-level stack interchange in Dallas.[1]

The $261 million[2] project began in 2002 and was completed in December 2005.[2]

High Five interchange.

Scope

The former interchange was a partial cloverleaf interchange ("parclo") modified for freeway speeds. The westbound lanes of I-635 were two levels above the eastbound lanes, while the intersecting US 75 was the mid-level of the interchange, crossing eastbound I-635 and then passing under westbound I-635.

As part of the project, I-635 was widened to include high-occupancy vehicle lanes ("HOV lanes"). The new lanes for regular traffic were built on the outer vacant right-of-way, while the HOV lanes were built in the median. The original I-635 lanes passing through the former interchange, now demolished, lay between the new regular and HOV lanes.

During construction staging, to maintain traffic flow, most of the flyover ramps were built over the existing parclo interchange. The old junction could then be demolished with little disruption once the new lanes were operational. Virtually none of the original junction remains after the reconstruction.

The interchange comprises 37 bridges distributed across 5 levels (thus the name "High Five"), with the highest ramps 120 feet (37 m) above ground.[3] The High Five consists of a total of 710 support tiers, 30,000 feet of drainage pipes and nearly 60 lane-miles of new roadway and stretches 2.4 miles (3.9 km) north/south and 3.4 miles (5.5 km) east/west.[1]

From top to bottom, the highways on the five levels are:

- I-635 exit ramps to US 75 and direct HOV connector, reversible for either EB I-635 to NB US 75 or SB US 75 to WB I-635
- US 75 exit ramps to I-635
- I-635 (LBJ Freeway)
- Frontage roads for US 75 and I-635
- US 75 (North Central Expressway)

The High Five also incorporated a hike and bike trail, named the Cottonwood Trail, which runs under all levels of the interchange. The section of the trail passing beneath the interchange was constructed as part of the High Five project by the Texas Department of Transportation.[4]

Awards

In 2006, the American Public Works Association named the High Five "a Public Works Project of the Year".[1]

References

[1] Dallas High Five Interchange Honored as Public Works Project of the Year (http://www.apwa.net/Documents/Advocacy/Press Release Archive/2006/06 June/6-26-06 Dallas High Five Interchange honored as Public Works Project of the Year.PDF). Retrieved 16 February 2009.

[2] TxDOT History, Present to 2001 (http://www.dot.state.tx.us/about_us/present_2001.htm). Texas Department of Transportation. Retrieved 11 August 2006.

[3] The Dallas High Five (http://www.zachry.com/projects_TXDot.htm). Zachry Construction Corporation. Retrieved 1 December 2006.

[4] City of Dallas Parks and Recreation. "Cottonwood Trail" (http://www.dallasparks.org/Downloads/Trails/cottonwood.pdf). . Retrieved 2010-02-23.

External links

- Google Maps Satellite View (http://maps.google.com/maps?f=q&source=s_q&hl=en&geocode=&q=high+five+interchange&sll=32.897966,-96.900616&sspn=0.041727,0.065746&ie=UTF8&hq=High+Five+Interchange&hnear=High+Five+Interchange,+Dallas,+TX+75243&ll=32.923114,-96.765046&spn=0.010429,0.024612&t=k&z=16)
- Texas Freeway, includes picture of the old interchange (http://www.texasfreeway.com/Dallas/construction/high_five/high_five_main.shtml)
- Alternate view of old interchange (http://www.texasfreeway.com/Dallas/photos/north_dallas_aerial/images/635_at_75_best_hres.jpg)
- Texas Transportation Institute (http://tti.tamu.edu/publications/researcher/newsletter.htm?vol=40&issue=2&article=7)
- Another view of the old interchange (http://tti.tamu.edu/publications/researcher/v40n2/images/high5before.jpg)

Interstate_635_(Texas)

WARNING: Article could not be rendered - ouputting plain text.

Potential causes of the problem are: (a) a bug in the pdf-writer software (b) problematic Mediawiki markup (c) table is too wide

Interstate 635Route information Maintained by Texas Department of TransportationTxDOTLength: 36.999 miTexas Department of Transportation, (59.5 km)Existed: 1 October 1959Oglesby, Scott. 3-digit Interstates from I-35. Kurumi. URL accessed 23 February 2005. – presentMajor junctionsSouth end: Interstate 20 (Texas)I-20 in Balch Springs, TexasBalch Springs Interstate 35E (Texas)I-35E in Dallas U.S. Route 75 in TexasUS 75 in Dallas Interstate 30 (Texas)I-30 in Mesquite, TexasMesquite U.S. Route 80 in TexasUS 80 in Mesquite, TexasMesquiteNorth end: Texas State Highway 121SH 121 near Dallas-Fort Worth International AirportDFWHighway systemAuxiliary route of the Interstate Highway SystemTexas state highwaysHighways in TexasInterstate 635 signage on the Dallas North Tollway frontage road. Galleria Dallas can be seen in the background.Interstate HighwayInterstate 635 or the Lyndon B. Johnson Freeway (abbreviated I-635 or locally, LBJ) is a partial loop around Dallas, Texas (United StatesUSA) between Interstate 20 (Texas)Interstate 20 in Balch Springs, Texas and State Highway 121 (Texas)State Highway 121 at the north entrance of the Dallas-Fort Worth International Airport in Grapevine, Texas. The route is 37 miles (60 km) long. It intersects Interstate 35E (Texas)Interstate 35E at Exits 27 B-C, but does not connect with Interstate 35W (Texas)Interstate 35W.HistoryA 1955 photo of the Interstate Highway city plans Interstate 635 was originally designated by the Texas Department of Transportation in 1959 as a loop around the eastern side of Dallas, connecting with I-35E to the northwest and southwest sides of the city. It was known locally as the Lyndon B. Johnson Freeway, or "LBJ" (named after the president). The section from Interstate 35E to U.S. Route 75 (Texas)U.S. Highway 75 (Central Expressway (Dallas)Central Expressway) was one of the busiest stretches of road in the Dallas-Fort Worth Metroplex, at virtually all hours of the day and night. It was a 10-mile (16 km) section and opened to motorists in January 1969, from US 75 to Barnes Bridge in Mesquite. The highway was extended westward to Interstate 35E northwest of Dallas, opening in February 1970. The designation was shortened in 1971 when Interstate 20 was rerouted south of Dallas, taking over 13 miles (21 km) of I-635's former route.http://www.dot.state.tx.us/tpp/hwy/ih/ih0635.htm TxDOT Highway Designation File: Interstate 635 The connecting section of I-20 from the west was not completed until 1978, at which time the western portion of I-635 was briefly concurrent with I-20 between the southern intersection with I-35 and US 80. When the outlet for I-20 from southeast of Dallas to Terrell was completed, the I-635 designation was truncated from I-35E to its intersection with I-20. An additional westward expansion (to the north entrance of DFW Airport) was designated in 1974 and opened in the 1980s. Because a portion of I-20 through Dallas County retains the name of Lyndon B. Johnson Freeway, together with the I-635 freeway, Dallas as what can be considered a partial (or 3/4) orbital road. The High Five InterchangeHigh Five construction project (so called because it is five levels and rises almost 120 feet (40 m) above the lowest level), a rebuild of the interchange of I-635 and US 75, was opened for traffic in February 2006. This interchange carries over 500,000 vehicles per day and was built as the largest interchange in the state of Texas to handle this vehicle load. A 2.7 billion United States dollardollar project was started on May 16, 2011 to widen I-635 and dig subsequent HOT LanesHOT lane tunnels beneath the primary roadway from I-35E to the High Five Interchange, a length of seventeen miles. Construction time is estimated at five years.Betz, Jonathan HOV lanes close on LBJ Freeway as 5-year, $2.7 billion makeover begins in North Dallas, Dallas Morning News, Retrieved May 16, 2011 The project is among several billion dollar plus projects in the planning phase in and around downtown Dallas (along with the rebuild of the I-35E & I-30 "Mixmaster" and the new Trinity Parkway). Route description Interstate 635 begins at an intersection with Interstate 20 in TexasI-20 in southeast Dallas, and travels northward through Balch Springs, TXBalch Springs into Mesquite, TXMesquite, where intersects US Route 80 and Interstate 30 in TexasI-30. The route then turns to the northwest, continuing near the border between Dallas and Garland, TXGarland. 6 miles (9.7 km) later is takes a general westward turn as it intersects U.S. Route 75 at the High Five Interchange. The route continues west, intersecting the Dallas North Tollway and its original terminus, Interstate 35E in Farmers Branch. The route then continues to the northwest, intersecting the President George Bush Turnpike in Irving, TXIrving before arriving at its final terminus at State Highway 121 (Texas)State Highway 121 at the north entrance to the Dallas-Fort Worth International Airport. Exit listCounty Location Exit number#Destinations Notes Dallas County, TexasDallasBalch Springs, TexasBalch Springs Interstate 20 (Texas)I-20 west – Fort Worth, TexasFort WorthSouthbound exit and northbound entrance 0 Interstate 20 (Texas)I-20 east – Tyler, TexasTyler, Shreveport, LouisianaShreveportSouthbound exit and northbound entrance 1A Seagoville Road 1B Elam Road 2 Lake June Road Mesquite, TexasMesquite3 West Cartwright Road, Bruton Road 4 Texas State Highway 352SH 352 (Scyene Road, Military Parkway) 5 Gross Road 6 U.S. Route 80 in TexasUS 80 – Dallas, TexasDallas, Terrell, TexasTerrellSigned as exits 6A (west) and 6B (east) 7A Towne Centre Drive Southbound exit and northbound entrance 7B Town East Boulevard – Town East MallSigned as exit 7 northbound 8 Interstate 30 (Texas)I-30 (U.S. Route 67US 67) – Dallas, TexasDallas, Texarkana, TexasTexarkanaSigned as exits 8A (west) and 8B (east) 9A Oates Drive, Galloway Avenue 9B La Prada Drive Southbound exit and northbound entrance Garland, TexasGarland11A Centerville Road, Ferguson Road 11B Northwest Highway, Shiloh Road Signed TO Texas State Highway 78SH 78 (Garland Avenue) westbound 12 Texas State Highway 78SH 78 (Garland Road) Eastbound exit and westbound entrance 13 Jupiter Road, Kingsley Road, Walnut Hill Lane Dallas14 Plano Road, Church Road 15 Miller Road, Royal Lane 16 Skillman Street, Audelia Road 17 Forest Lane, Abrams Road 18A Greenville Avenue 18B TI Boulevard Westbound exit is via exit 18A 19A U.S. Route 75 in TexasUS 75 (Central Expressway (Dallas)Central Expressway) – Dallas, TexasDallas, McKinney, TexasMcKinney19B Coit Road 20A Park Central Drive Westbound exit and eastbound entrance 20B Hillcrest Road Signed as exit 20 eastbound 21 Texas State Highway 289SH 289 (Preston Road) 22A Montfort Drive Westbound exit and eastbound entrance 22B Dallas Parkway, Inwood Road Eastbound exit is via exit 22D 22C-D Dallas North TollwaySigned as exits 22C (north) and 22D (south) 23 Midway Road, Welch Road 24 Marsh Lane 25 Webb Chapel Road 26 Josey Lane Westbound exit is via exit 25 27A Denton Drive, Harry Hines BoulevardWestbound exit and eastbound entrance 27B-C Interstate 35E (Texas)I-35E (Stemmons Freeway) – Denton, TexasDenton, Dallas, TexasDallasSigned as exits 27B (north) and 27C (south), left lane exit Farmers Branch, TexasFarmers Branch28 Luna Road Irving, TexasIrving29 Valley View Lane Westbound exit and eastbound entrance 29 President George Bush Turnpike/Las Colinas Boulevard, Valley View Lane Eastbound exit and westbound entrance; signed as exits 29A (south) and 29B (north) 30 President George Bush Turnpike south Westbound exit and eastbound entrance 31 MacArthur Boulevard 31B Olympus Boulevard Eastbound exit is via exit 31 33 Belt Line Road (Dallas)Belt Line Road34 Freeport Parkway Coppell, TexasCoppell35 Royal Lane Tarrant County, TexasTarrantGrapevine, TexasGrapevine36A Texas State Highway 121SH 121 south – Dallas/Fort Worth International AirportD/FW AirportWestbound exit and eastbound entrance 36B Texas State Highway 121SH 121 north / Bass Pro Drive Westbound exit and eastbound entrance See alsoList of facilities named after Lyndon JohnsonReferencesExternal links Interstate Guide: I-635 (Texas) I-635 info page -- from dfwfreeways.info

Central_Expressway_(Dallas)

WARNING: Article could not be rendered - ouputting plain text.

Potential causes of the problem are: (a) a bug in the pdf-writer software (b) problematic Mediawiki markup (c) table is too wide

Central Expressway at Fitzhugh AvenueCentral Expressway is a north–south highway in Dallas, TexasDallas, Texas (United StatesUSA) and surrounding areas.North Central Expressway The best-known section is the North Central Expressway, a name for a freeway section of U.S. Highway 75 (Texas)U.S. Highway 75 between downtown Dallas and McKinney, Texas. The southern terminus is at an intersection with "hidden" Interstate 345 (Texas)Interstate 345 (signed as Interstate 45 (Texas)Interstate 45) and Spur 366 (Texas)Woodall Rodgers Expressway. www.interstate-guide.comSouth of US 75's terminus, North Central Expressway briefly continues south in the median (road)median of I-345, then becomes a surface street through the eastern side of downtown Dallas. The surface street section south of Pearl Street was renamed for César Chávez in April, 2010 Cesar Chavez Street Signs Debuted Today At City Hall, Along Cesar Chavez Blvd., Unfair Park blog, Dallas Observer, Patrick Michels--writer, April 9, 2010. Retrieved 2010-04-10.. The North Central Expressway is near high-income neighborhoods and enclave cities such as Highland Park, TexasHighland Park and University Park, TexasUniversity Park. The freeway is also adjacent to popular districts including Uptown DallasUptown, Cityplace, DallasCityplace, Lower Greenville, Dallas, TexasLower Greenville, NorthPark Center, and the Telecom Corridor. Near the intersection of Central Expressway and Mockingbird Lane is Southern Methodist University, and Mockingbird Station. The Dallas Area Rapid Transit light rail system has a tunnel underneath the North Central Expressway between downtown Dallas and Mockingbird Station. History The Central project was first proposed by Dallas City Planner George E. Kessler in 1911, who suggested that the city buy the right of way of the Houston and Texas Central Railroad (H&TC) to remove the railway tracks and construct a Central Boulevard (later renamed the Central Expressway project) in their place. The Central project became a real project in the 1920's (with first mention in The Dallas Morning News in December 1924), but resistance from the Southern Pacific RailroadSouthern Pacific railroad company proved to be a serious obstacle that delayed the project for decades. Southern Pacific RailroadSouthern Pacific opposed the use of their railroad's right of way to construct the Central Expressway, and it was this opposition and lobbying of political forces that caused the significant delays in the construction to the early 1950s. VickeryPlace.com: archive of newspaper clippings on the History of Central ExpresswayParts of the North Central Expressway were opened in 1950. The route from Downtown to Mockingbird Lane was fully functional by the end of 1952, and the whole route to Campbell Road in Richardson, TexasRichardson was opened for traffic in August 1956. By the time the Central Expressway opened for traffic, North Dallas and Richardson had already expanded beyond expectations, and the new highway was already hopelessly inadequate by the 1960s. The Expressway did not reach the city of Anna, TexasAnna and the northern Collin County, TexasCollin County line until late 1969 or early 1970. Reconstruction Prior to reconstruction, the North Central Expressway was considered to be one of the most poorly designed freeways in the nation. Texasfreeway.com - Photos from after renovation Texasfreeway.com - Historic photos Though initially an engineering marvel as Dallas's first freeway when it opened to traffic in 1950, the explosive growth that soon hit north Dallas and the nearby suburbs quickly overwhelmed its design and capacity. By the 1980s, the four-lane freeway had acquired a reputation for severe rush-hour traffic jams. In the early 1980s, the Texas Department of TransportationTxDOT floated plans to build an elevated structure above the existing freeway. After considerable study and debate, elevated structures were eliminated. Construction started in 1992 and was finally completed in November 1999. Total reconstruction cost was around United States Dollar$600 million.

Current state The facility has 8 continuous general-purpose lanes from downtown Dallas to Legacy Drive in Plano except for a six-lane segment where it passes under Interstate 635 (Texas)Interstate 635 (two additional lanes are present but are only entrance ramps/exit ramps for Forest Lane and Midpark Road). For the six miles (10 km) north of downtown Dallas, the freeway lies more than 30 feet (9.1 m) below adjacent and partially cantilevered frontage roads. Aesthetic treatment The freeway's architecturally distinctive design distinguishes it as one of the nation's most attractive urban freeways. Every structure and element along the highway right-of-way was given aesthetic attention during the design phase. Support columns for overpasses and bridges have been designed to be as visually appealing as possible. The beige concrete columns which form the support structure for the retaining walls contrast with the brown, textured infill panels of the walls to create a multicolored and articulated edge to the freeway. Two million square feet of these walls along the nine mile (14 km)-long project distinguishes the freeway. Freedman's Cemetery Central Expressway's 1940s construction methods would come to represent inadequacies of both civil engineering and social engineering. The southern end of North Central Expressway passed through a historic African-American neighborhood, displacing 1,500 black residents. But the full scale of the injustice was not revealed until preparations began for the 1990s expansion of the route. The living weren't the only ones displaced; a quarter of the 4-acre (16000 m2) Freedman's Cemetery, with graves dating back to Emancipation ProclamationEmancipation, had been paved over. Archeological excavations uncovered the remains of over 1,100 men, women, and children under existing and proposed roadways. After their reburial, the site was turned into a memorial to the working-class black residents of the area, which after the expressways were built, became the upscale Uptown Dallas neighborhood.http://www.projectpast.org/jcbrandon/papers/dallas2004.aspSouth Central Expressway From south of Main Street, DallasMain Street and its crossing under the Interstate 45 (Texas)Interstate 45 overhead in downtown Dallas, Central Expressway became the South Central Expressway. The road met Interstate 30 (Texas)Interstate 30 at a three-level interchange and continued south as a surface road down to I-45. This section was renamed in honor of César Chávez in April, 2010 Cesar Chavez Street Signs Debuted Today At City Hall, Along Cesar Chavez Blvd., Unfair Park blog, Dallas Observer, Patrick Michels--writer, April 9, 2010. Retrieved 2010-04-10.. Upon crossing under I-45, it becomes U.S. Route 175U.S. Highway 175 and is signed both as South Central Expressway and S.M. Wright Freeway (named for a local minister). Where US 175 turns off to the east, the road continues south as Texas State Highway 310State Highway 310; the S.M. Wright Freeway name ends at Loop 12 (Texas)Loop 12, past the end of freeway standards. The South Central Expressway continues past a partial interchange with Interstate 20 (Texas)Interstate 20 to end at a merge with I-45 north of Hutchins, TexasHutchins. This section through and south of downtown mainly serves local traffic. I-45 was built roughly parallel to it. However, the parallel section of I-45 north of the Trinity River is completely elevated, and is often difficult to traverse when ice storms hit Dallas (about once per year); thus, South Central often takes over some of the traffic during this time.Future changes The US 175 section is due to be rerouted in the future, off South Central/S.M. Wright onto a new alignment, further extending C.F. Hawn Freeway west to I-45; it has not been reported whether or not the State Highway 310 numbering would replace US 175's when it is rerouted. Meanwhile, TxDOT is preparing to redesign the portion of South Central/S.M. Wright (between I-45 and US 175's current east turn) into a surface-street boulevard, and has held meetings with local residents about the project. Many have differences of opinion; some want a 4-lane arrangement, others want 6 lanes. Some even think TxDOT is to blame for dividing the surrounding neighborhood. Unfortunately, there is not current funding for the project, and even if there were, it would be 2015 at the earliest before US 175's rerouting could be finished TxDOT Dazzles Southern Dallas With Plans to Straighten Out "Statistically Dangerous Curve" on S.M. Wright, Dallas Observer "Unfair Park" Blog, Daniel Rodrigue--writer, March 31, 2010. Retrieved 2010-05-22. SMWrightProject.org, Halff Associates, Inc. Retrieved 2010-05-22 Some say fix for S.M. Wright Freeway isn't enough, Dallas Morning News, Michael A. Lindenburger--writer, May 16, 2010. Retrieved 2010-05-22.. Exit list South Central Expressway County Location Destinations Notes Dallas County, TexasDallasDallas, TexasDallasLamar Street northbound exit and southbound entrance Municipal Avenue; Lamar Street northbound exit and entrance Haven Street southbound exit and entrance U.S. Highway 175 (Texas)U.S. Highway 175 east - Kaufman (TX)Kaufmanno southbound entrance Hatcher Street no northbound exit Pine Street Metropolitan Avenue

Pennsylvania Avenue northbound exit and southbound entrance Martin Luther King Jr. Boulevard northbound exit only Interstate Highway 45 to U.S. Highway 75 (Texas)U.S. Highway 75 - McKinney (TX)McKinneynorthbound exit and southbound entrance Good-Latimer Expresswaynorthbound exit and southbound entrance At this point South Central Expressway becomes a surface road through Downtown Dallas. North Central Expressway North Central Expressway is the only highway in Texas to use a consecutive-numbering system as opposed to a mile marker system. The numbering system south of I-635 changed drastically when several exits were consolidated; north of I-635 the system remained the same. When US 75 was rerouted west of Sherman and Denison, additional exits took the next available numbers. North of McKinney, US 75 is not referred to as Central Expressway, giving way to other names such as Collins Freeway (Howe) and Sam Rayburn Freeway (Sherman). County Location exit number#Destinations Notes Dallas County, TexasDallasDallasOld Bryan Street; Good-Latimer Expresswaysouthbound exit and northbound entrance Interstate Highway 45 (Texas)Interstate Highway 45 south - Houston (TX)Houston (Interstate Highway 345 (Texas)Interstate Highway 345 south) southbound exit and northbound entrance 285 Ross Avenue southbound exit only; exit no. corresponds to I-45 1A to Interstate Highway 35E (Texas)Interstate Highway 35E - Denton (TX)Denton; Waco (TX)Waco (Spur 366 (Texas)Spur 366 west) southbound exit and northbound entrance; northbound where I-45 becomes US 75, exit 286; no. corresponds to I-45 north 1 (Hall Street) 2 (Lemmon Avenue) 3 1B Haskell Avenue; Blackburn Street; Fitzhugh Avenue; Lemmon Avenue 4 (Fitzhugh Avenue) 5 2 Knox Street; Henderson Avenue; Monticello Avenue; Fitzhugh Avenue 6 (Monticello Avenue) 7 (McCommas Boulevard) 8 3 Mockingbird Lane; University Boulevard; Monticello Avenue 9 (Yale Boulevard) 10 (University Boulevard) 11 4A Lovers Lane; Southwestern Boulevard; University Boulevard exit 4 southbound 12 (Southwestern Boulevard) 13 4B Caruth Haven Lane southbound exit is combined with 5A 14-15 5A Loop 12 (Texas)Loop 12 - Caruth Haven Lane 16 5B Northpark Boulevard; Park Lane 17 6 Walnut Hill Lane; Meadow Road 18 (Meadow Road) 19 7 Royal Lane; Meadow Road 20A 8A Forest Lane exit 20-20A southbound 20B 8B Coit Road northbound exit and southbound entrance 20B Frontage Road northbound exit only 21 Interstate Highway 635 (Texas)Interstate Highway 63520A Churchill Way southbound exit and northbound entrance 22 Midpark Road Richardson, TexasRichardson23 Spring Valley Road 24 Belt Line Road (Dallas)Belt Line Road; Main Street 25 Arapaho Road 26 Campbell Road northbound exit and southbound entrance 27A Galatyn Parkway; Renner Road; Campbell Road exit 26 southbound Collin County, TexasCollin27 Renner Road southbound exit and northbound entrance 28A Plano Parkway 28B President George Bush TurnpikePlano, TexasPlano29 15th Street (former Farm to Market RoadFM 544) northbound exit and southbound entrance 29A Park Boulevard; 15th Street 30 Parker Road 30A Frontage Road southbound exit only 31 Spring Creek Parkway 32 Legacy Drive 32A Frontage Road northbound exit and southbound exit Allen, TexasAllen33 Bethany Drive 34 McDermott Drive (former Farm to Market RoadFM 2170) 35 Allen Drive northbound exit and southbound entrance 36 Exchange Parkway 37 Farm to Market Road 2786 - Stacy Road 38 Ridgeview Drive 38A State Highway 121 (Texas)State Highway 121 south; to State Highway 5 (Texas)State Highway 5 - Fort Worth (TX)Fort Worth (Spur 399 (Texas)Spur 399; former Farm to Market Road 720) SH 121 joins northbound and leaves southbound McKinney, TexasMcKinney39 Eldorado Parkway 40A Farm to Market Road 3038; Spur 359 (Texas)Spur 359 - Louisiana Street; Virginia Parkway 40B White Avenue northbound exit and southbound entrance 41 U.S. Highway 380 (Texas)U.S. Highway 380 - Greenville (TX)Greenville; Denton (TX)Denton42 Wilmeth Road 43 Farm to Market Road 543 - Weston Road 44 Telephone Road; Davis Road 45 State Highway 121 (Texas)State Highway 121 north - Bonham (TX)BonhamSH 121 joins southbound and leaves northbound Melissa, TexasMelissa46 Melissa Road 47 Throckmorton Road; Foster Crossing Road Anna, TexasAnna48 Farm to Market Road 455 - Anna-Weston Road 49 Mantua Road 50 County Line Road Grayson County, TexasGraysonVan Alstyne, TexasVan Alstyne51 Farm to Market Road 121 - Van Alstyne (TX)Van Alstynedue to confusion between this exit and State Highway 121 (a few miles to the south, which leads to Dallas-Fort Worth International AirportDFW Airport), southbound highway signs are marked "DFW TRAFFIC DO NOT EXIT" 52 Farmington Road Howe, TexasHowe53 Spur 381 (Texas)Spur 381 - Howe (TX)Howe54 Farm to Market Road 902; State Highway 5 (Texas)State Highway 5Sherman, TexasSherman55A Frontage Road southbound exit and entrance 55 Frontage Road 56 Farm to Market Road

141756A South Travis Street northbound exit and southbound entrance 56B Frontage Road northbound exit only 57 Park Street; Center Street 58 Texas State Highway 56State Highway 56 - Lamar Street; Houston Street Lamar Street is one way eastbound; Houston Street is one way westbound 59 Washington Street no southbound entrance 60 Farm to Market Road 131 - Travis Street; Taylor Street 61 State Highway 91 (Texas)State Highway 91 north — Texoma Parkway no southbound exit 62 Lamberth Road northbound exit and southbound entrance 63 U.S. Highway 82 (Texas)U.S. Highway 82 - Bonham, TexasBonham; Gainesville, TexasGainesville64 Loy Lake Road; Fallon Drive 65 Farm to Market Road 691 - Grayson County Airport; Grayson County Junior CollegeDenison, TexasDenison66 Spur 503 (Texas)Spur 503 - Denison (TX)Denison67 Loy Lake Road 68 Crawford Street Road 69 Farm to Market Road 120 - Pottsboro (TX)Pottsboro; Lake Texoma70 Farm to Market Road 8471 Randell Lake Road 72 State Highway 91 (Texas)State Highway 9173 U.S. Highway 69 (Texas)U.S. Highway 69 south - Greenville (TX)Greenville; Denison (TX)DenisonUS 69 joins northbound and leaves southbound 74 Texas Travel Info Center Images of Central Expressway Central Expressway heading southbound, towards Downtown DallasCentral Expressway near NorthPark CenterCentral Expressway near Southern Methodist University and Mockingbird LaneCentral Expressway at Haskell AveReferences

Red_Line_(Dallas_Area_Rapid_Transit)

Red Line	

Northbound Red Line train at Mockingbird Center	
Overview	
Type	light rail
System	Dallas Area Rapid Transit
Status	operational
Locale	Dallas, Texas
Termini	Parker Road (north) Westmoreland (south)
Stations	25
Operation	
Opened	June 1996
Owner	DART
Operator(s)	DART
Technical	
Track gauge	4 ft 8 $^1/_2$ in (1435 mm)
Electrification	overhead catenary
Highest elevation	at grade, elevated, underground

The **Red Line** is a light rail line in the system of mass transit in Dallas, Texas (USA) operated by the Dallas Area Rapid Transit system. Along with the Blue Line, it is one of the original modern rail lines in Dallas.

The southwestern terminus of the line is at Westmoreland Station in southwest Dallas at the intersection of Illinois Avenue and Westmoreland Road. The line runs northeast, under the Dallas Convention Center and through downtown Dallas. Then the line follows Central Expressway (US 75) through north Dallas, Richardson and Plano, finally ending at the Parker Road Station at Park Boulevard near Central Expressway in Collin County.

The rail line was part of the initial launch of DART's light rail service in 1996. At the time, the line only ran from Westmoreland Station to Pearl Station in the northeast corner of downtown. In 1997, the Red Line was extended to Park Lane Station. On 18 December 2000, Cityplace Station, the southwest's first commercial subway station was opened along the Red Line underneath Cityplace Tower just northeast of downtown.

In 2002, the Red Line extended into Richardson, ending at Galatyn Park Station, extending the light rail service 9 miles (14.5 km) over its original length. Later that year, the line was opened to the Parker Road Station, an additional 3 miles (4.8 km) of track, its current terminus.

Stations

Daily Service

Listed from north to south

Station	Transfers	Station Type	Doors Open	Location	
Parker Road	Orange Line*	Surface	Left	Plano	
Downtown Plano	Orange Line*	Surface	Right		
Bush Turnpike	Orange Line*	Surface	Right	Richardson	
Galatyn Park	Orange Line*	Surface	Right		
Arapaho Center	Orange Line*	Surface	Right		
Spring Valley	Orange Line*	Elevated	Left		
LBJ/Central	Orange Line*	Surface	Right	North Dallas	Dallas
Forest Lane	Orange Line*	Elevated	Left		
Walnut Hill	Orange Line*	Elevated	Left		
Park Lane	Orange Line*	Elevated	Left	Vickery Meadow	
Lovers Lane	Orange Line*	Surface	Right	East Dallas	
Mockingbird	Blue Line, Orange Line*	Recessed	Left		
Cityplace	Blue Line, Orange Line*	Underground	Left	Cityplace, Oak Lawn	
Pearl	Blue Line, Green Line, Orange Line	Surface	Left	City Center District	
St. Paul	Blue Line, Green Line, Orange Line	Surface	Right		
Akard	Blue Line, Green Line, Orange Line	Surface	Right		
West End	Blue Line, Green Line, Orange Line	Surface	Right	West End Historic District	
Union Station	Blue Line, Trinity Railway Express, Amtrak	Surface	Right	Reunion District	
Convention Center	Blue Line	Surface	Right	Convention Center District	
Cedars	Blue Line	Surface	Right	Cedars, South Dallas	
8th & Corinth	Blue Line	Surface	Right	Oak Cliff	
Dallas Zoo		Surface	Right		
Tyler/Vernon		Surface	Right		
Hampton		Surface	Right		
Westmoreland		Surface	Either		

* Orange Line only serves Pearl–Parker Road during the morning and evening rush hours.

Special Event Service

- Deep Ellum Station (Served regularly by Green Line, Orange Line)
- Baylor University Medical Center Station (Served regularly by Blue Line, Orange Line)
- Fair Park Station (Served regularly by Green Line, Orange Line)
- Victory Station (Also served by Blue Line, Green Line, Orange Line)

Cedars Station in the Cedars neighborhood of south Dallas

Union Station in the Reunion District of downtown

West End Station in the West End Historic District of downtown

External links

- Full DART System Map [1]

Orange_Line_(Dallas_Area_Rapid_Transit)

Orange Line	
Overview	
System	light rail
Operation	
Opened	December 6, 2010
Technical	
Track length	14 miles
Track gauge	4 ft 8 $^1/_2$ in (1435 mm)

The **Orange Line** (labeled as the **Purple Line** on maps prior to 2006) will be a 14 mile (23 km) mass transit light rail line in Dallas, Texas (USA) operated by the Dallas Area Rapid Transit system. It will operate in addition to the currently existing Blue, Red and Green Line lines. The Orange Line started operation on December 6, 2010 with weekday peak service from the Parker Road Station to Bachman Station. The Orange Line shares alignment with the Green Line until its split north of Bachman Station. It was originally scheduled to open in phases starting in December 2011, with a route through downtown Dallas terminating at DFW Airport Station by December 2014. Openings of these phases, however, have been delayed with the plans and timetables subjected to many updates. Through the use of interlining with Red Line and Green Line tracks, the Orange Line will terminate at Lawnview Station. During peak hours, Orange line trains will also terminate at Parker Road Station.[1] This peak hour service was initiated on December 6, 2010 in conjunction with the opening of the Green Line.

Construction of the Irving Convention Center near the future Irving Convention Center Station

Future phases of the Orange line include a DFW North Station loop[2] and an eastern extension down Scyene Road to Masters Drive (previously planned as a Green Line expansion).[3]

Planned opening

Opened December 2010

The following stations began selected weekday trips in conjunction with the opening of the Green Line:

Listed from East to West

- Parker Road (Also served by Red Line)
- Downtown Plano (Also served by Red Line)
- Bush Turnpike (Also served by Red Line)
- Galatyn Park (Also served by Red Line)
- Arapaho Center (Also served by Red Line)
- Spring Valley (Also served by Red Line)
- LBJ/Central (Also served by Red Line)
- Forest Lane (Also served by Red Line)
- Walnut Hill (Also served by Red Line)
- Park Lane (Also served by Red Line)
- Lovers Lane (Also served by Red Line)
- Mockingbird (Also served by Red Line and Blue Line)
- Cityplace (Also served by Red Line and Blue Line)
- Pearl Station (Also served by Red Line, Blue Line, and Green Line)
- St. Paul Station (Also served by Red Line, Blue Line, and Green Line)
- Akard Station (Also served by Red Line, Blue Line, and Green Line)
- West End Station (Also served by Red Line, Blue Line, and Green Line)
- Victory Station (Also served by Green Line and Trinity Railway Express)
- Market Center Station (Also served by Green Line)
- Southwestern Medical District/Parkland Station (Also served by Green Line)
- Inwood/Love Field Station (Also served by Green Line)
- Burbank Station (Also served by Green Line)
- Bachman Station (Also served by Green Line)

Opening December 2011

Listed from East to West

- Lawnview Station (Also served by Green Line)
- Hatcher Station (Also served by Green Line)
- MLK, Jr. Station (Also served by Green Line)
- Fair Park Station (Also served by Green Line)
- Baylor University Medical Center Station (Also served by Green Line)
- Deep Ellum Station (Also served by Green Line)

Opening August 2012

- University of Dallas Station
- Las Colinas Urban Center Station
- Irving Convention Center Station
- North Lake College Station
- Belt Line Station

Opening December 2014

- DFW Airport Station[4]

Future/Deferred

- Loop 12 Station (Deferred)
- South Las Colinas Station (Deferred)
- Las Colinas Carpenter Ranch Station (Deferred)[5]
- DFW North Station (Future branch)

Construction issues

On 12 March 2007, the City of Dallas officials and DART made an agreement to make Love Field Station a surface-level facility after a long debate of whether or not to make it an underground station.[6] [7]

On December 5, 2007, the *Dallas Morning News* ran a story reporting that DART President Gary Thomas stated that a previous cost estimate of $988 million was too low. A new cost estimate for the 14 mile project was $1.8 - $1.9 billion.[8] The $900 million overrun in costs has caused considerable outrage amongst political leaders[9] in the City of Irving Texas, where the line passes on its way to Dallas Fort Worth International Airport. Irving leaders subsequently made inquiry as to the cost overruns.[10] Texas State Representative Linda Harper Brown sent an official letter [11] to Mr. Thomas also making inquiry into the project's substantial cost overruns.

In February 2010 DART officials announced that the first two phases may be delayed due to TXDOT problems along State Highway 114, which the Orange Line route follows. Utility relocation and road construction will delay access to portions of the construction area where the rail line and highway intersect. The delay could push the connection to Las Colinas from December 2011 to August 2012. DART is determined to keep the original schedule and minimize any delay.[12] [13]

In June 2010, DART placed the line on indefinite hold due to declining revenue; on 15 September 2010, the agency said that due to cost savings and federal funds, the plans for the line have been revived.[14]

References

[1] http://www.dart.org/about/expansion/noiorangelinedfw.pdf

[2] http://www.dart.org/about/expansion/orangelinedfwnewsletter1.pdf

[3] http://www.dart.org/about/expansion/2030plan.asp

[4] http://www.dallasnews.com/sharedcontent/dws/news/localnews/stories/091510dnmetdartbudget.f1c2685f.html

[5] http://www.dart.org/about/expansion/orangelinestations.asp

[6] *The Dallas Morning News* (http://www.dallasnews.com) - *9 February 2007.*

[7] " Rail will not tunnel under Love Field (http://www.dallasnews.com/sharedcontent/dws/news/localnews/stories/031307dnmetdartlove.39c0619.html)." Retrieved 12 March 2007

[8] A new cost estimate (http://www.dallasnews.com/sharedcontent/dws/news/localnews/stories/120507dnmetdart.18cd6b.htm)

[9] outrage amongst political leaders (http://www.dallasnews.com/sharedcontent/dws/dn/latestnews/stories/120807dnmetirvdevelop.226610b.html)

[10] inquiry (http://www.dallasnews.com/sharedcontent/dws/dn/latestnews/stories/121307dnmetIrving.c724414.html)

[11] an official letter (http://www.dallasnews.com/sharedcontent/dws/img/12-07/1212dart.pdf)

[12] http://www.dallasnews.com/sharedcontent/dws/dn/latestnews/stories/DN-dartside_24met.ART.State.Edition2.4bab989.html "Irving light-rail segments may be delayed." The Dallas Morning News. February 24, 2010.
[13] http://www.dart.org/news/news.asp?ID=898 DART
[14] "DART finance committee OKs service changes, new debt in $1.26 billion budget" (http://www.dallasnews.com/sharedcontent/dws/news/localnews/stories/091510dnmetdartbudget.f1c2685f.html). 15 September 2010. . Retrieved 15 September 2010.

External links

- Full DART System Map (http://www.dart.org/maps.asp?zeon=RailMap)
- DART Expansion Plans (http://www.dart.org/mis.asp?zeon=DARTRailExpansionFacts)
- DART Orange Line Expansion Information (http://www.dart.org/about/expansion/orangeline.asp)

Forest_Lane_(DART_station)

Forest Lane DART light rail station	

Station statistics	
Address	8210 Forest Lane Dallas, Texas, 75243
Coordinates	32°54′31″N 96°45′42″W
Lines	Red Line Orange Line
Connections	DART Routes 234 [1] (M-F), 360 [1], 451 [2], 486 [3], 488 [4], and 567 [2] Medical City E-Shuttle (M-F)
Platforms	Island
Parking	271 free spaces, no overnight
Bicycle facilities	Two lockers, one rack
Other information	
Opened	1 July 2002
Accessible	♿
Owned by	Dallas Area Rapid Transit
Services	

Preceding station	Dallas Area Rapid Transit	Following station
Walnut Hill *toward Westmoreland*	Red Line	LBJ/Central *toward Parker Road*
Walnut Hill *toward Bachman*	Orange Line	
	Effective Late 2011	
Walnut Hill *toward Irving Convention Center*	Orange Line Parker Road branch	LBJ/Central *toward Parker Road*

Forest Lane Station is a DART Light Rail station located in north Dallas, Texas (USA) on Forest Lane, just east of 75 US 75 (North Central Expressway). It opened on July 1, 2002 and is a station on the Red Line, serving nearby residential and shopping areas.

External links

- DART - Forest Lane Station [5]

Westmoreland_(DART_station)

Westmoreland **DART light rail station**	

Station statistics	
Address	2646 South Westmoreland Road Dallas, Texas 75211
Coordinates	32°43′11″N 96°52′18.5″W
Lines	Red Line
Connections	DART Routes 404 [1] (M-F), 445 [2], 547 [3] (M-Sat), 548 [4], 549 [5], 550 [6], 568 [7], and 574 [8] (M-F)
Platforms	Island platform
Parking	693 free spaces, no overnight
Bicycle facilities	Two racks, two lockers
Other information	
Opened	June 14, 1996[1]
Accessible	♿
Owned by	Dallas Area Rapid Transit
Services	

Preceding station	Dallas Area Rapid Transit	Following station
Terminus	Red Line	Hampton *toward Parker Road*

Westmoreland is a DART Light Rail station located in the Oak Cliff neighborhood of Dallas, Texas (USA) at Illinois Avenue and Westmoreland Road. It opened on June 14, 1996 and is the southwest terminus of the Red Line, serving the nearby residential and commercial areas and featuring bus service to Southwest Center Mall.

References

[1] Howell, Curtis; Mark Wrolstad (June 15, 1996). "Now Boarding: Rail debut has crowds lining up". *The Dallas Morning News*.

External links

- DART - Westmoreland Station (http://www.dart.org/riding/stations/westmorelandstation.asp)

Bachman_(DART_station)

Bachman Station

Station statistics	
Address	South of Northwest Hwy. on the west side of Denton Drive
Coordinates	32°51′15″N 96°52′38″W
Lines	Green Line Orange Line
Connections	DART Routes 31 [1], 428 [2], 528 [3], 535 [4], 544 [5]
Structure	At Grade
Platforms	2 island platforms
Parking	443 spaces
Other information	
Opened	December 6, 2010[1]
Accessible	♿
Owned by	Dallas Area Rapid Transit

Services

Preceding station	Dallas Area Rapid Transit	Following station
Walnut Hill/Denton *toward North Carrollton/Frankford*	Green Line	Burbank *toward Buckner*
Terminus	Orange Line	Burbank *toward Parker Road*
	Effective Late 2011	
University of Dallas *toward Irving Convention Center*	Orange Line	Burbank *toward Lawnview or Parker Road*
	Deferred	
Loop 12 *toward Irving Convention Center*	Orange Line	Burbank *toward Lawnview or Parker Road*

Bachman Station is a DART light rail station located in Northwest Dallas for service on the Green Line and Orange Line.[2] [3] The station opened as part of the Green Line's expansion in December 2010. It will serve nearby attractions such as Bachman Lake and surrounding neighborhoods. It will be a major transfer station between the Green and Orange lines.

References

[1] Lindenberger, Michael A. (December 5, 2010). "New Green Line 'starts to complete' DART vision" (http://www.dallasnews.com/sharedcontent/dws/dn/latestnews/stories/120510dnmetgreenmain.1bf8a3c.html). *The Dallas Morning News*. . Retrieved December 7, 2010.

[2] DART Green Line Expansion Information (http://www.dart.org/about/expansion/greenline.asp).

[3] DART Orange Line Expansion Information (http://www.dart.org/about/expansion/orangeline.asp).

External links

- Dallas Area Rapid Transit (http://www.dart.org)

Irving_Convention_Center_(DART_station)

Irving Convention Center Station	
Station statistics	
Coordinates	32°52′36″N 96°56′17″W
Lines	Orange Line
Structure	At Grade
Other information	
Opened	*Proposed to open July 30, 2012*
Accessible	♿
Owned by	Dallas Area Rapid Transit
Services	

Preceding station	Dallas Area Rapid Transit	Following station
	Effective Late 2011	
Terminus	Orange Line	Las Colinas Urban Center *toward Lawnview or Parker Road*
	2012 or later	
North Lake College *toward DFW Airport*	Orange Line	Las Colinas Urban Center *toward Lawnview or Parker Road*
	Deferred	
Las Colinas Carpenter Ranch *toward DFW Airport*	Orange Line	Las Colinas Urban Center *toward Lawnview or Parker Road*

Irving Convention Center Station will be a DART light rail station located in the Las Colinas development of Irving, Texas for service on the Orange Line.[1] The station is planned to open in December 2011 and will serve the Irving Convention Center and Las Colinas Urban Center. This station will temporarily serve as the Orange Line terminus until expansion in December 2012.

See also

- North Irving Transit Center

References

[1] DART Orange Line Expansion Information (http://www.dart.org/about/expansion/orangeline.asp).

External links

- Dallas Area Rapid Transit (http://www.dart.org)

Spring_Valley_(DART_station)

Spring Valley DART light rail station	

Station statistics	
Address	100 West Spring Valley Road Richardson, Texas, 75081
Coordinates	32°56′27″N 96°44′15″W
Lines	Red Line Orange Line
Connections	DART Routes 360 [1], 400 [1], 463 [2], and 571 [3] (M-F)
Structure	Elevated
Platforms	Island platform
Parking	403 free spaces, no overnight
Bicycle facilities	two lockers, one rack
Other information	
Opened	1 July 2002
Accessible	♿
Owned by	Dallas Area Rapid Transit
Services	

Preceding station	Dallas Area Rapid Transit	Following station
LBJ/Central *toward Westmoreland*	Red Line	Arapaho Center *toward Parker Road*
LBJ/Central *toward Bachman*	Orange Line	
	Effective Late 2011	
LBJ/Central *toward Irving Convention Center*	Orange Line Parker Road branch	Arapaho Center *toward Parker Road*

Spring Valley Station is a DART Light Rail station located in Richardson, Texas (USA) off Spring Valley Road, just east of 75 US 75 (Central Expressway). It opened 1 July 2002 and is a station on the Red Line, serving nearby residential areas and the nearby Blue Cross/Blue Shield of Texas.

The station provides both bus service and DART light rail service, along with over 400 free parking spaces ("Park and Ride"). In 2008, construction began just east of the station of a combination residential / retail development "Spring Valley Station District", a planned development district within the City of Richardson.

External links

- DART - Spring Valley Station [4]
- Spring Valley Station District [5]

Parker_Road_(DART_station)

Parker Road DART light rail station	
 The main tracks and platforms at Parker Road Station.	
Station statistics	
Address	805 East Park Boulevard Plano, Texas 75074
Coordinates	33°2′2″N 96°42′3″W
Lines	Red Line Orange Line
Connections	DART Routes 350 [1] (M-Sat), 410 [2], and 452 [3] DART TI Spring Creek Shuttle [4] (M-F) DART On-Call East Plano [5] (M-F) DART On-Call North Central Plano [6] (M-F) Texoma Express (To/from Sherman) (M-F)
Platforms	Island
Parking	2,078 free spaces, no overnight
Other information	
Opened	December 9, 2002
Accessible	♿
Owned by	Dallas Area Rapid Transit
Services	

Preceding station	Dallas Area Rapid Transit	Following station
Downtown Plano *toward Westmoreland*	Red Line	*Terminus*
Downtown Plano *toward Bachman*	Orange Line	
	Effective Late 2011	
Downtown Plano *toward Irving Convention Center*	Orange Line Parker Road branch	*Terminus*

Parker Road Station is a DART Light Rail station located in Plano, Texas (US) in Collin County at the intersection of Park Boulevard and Archerwood Street near 75 US 75 (Central Expressway). It opened December 9, 2002, and is the northern terminus of the Red Line. It was originally called **East Plano Transit Center** before the light rail station opened up. The Orange Line terminates some trains at Parker Road Station during peak hours for direct service to DFW Airport Station.[1] In July of 2011, it was announced that Parker Road Station would be part of a pilot program that charged passengers from non-DART cities for parking.[2]

References

[1] http://www.dart.org/about/expansion/noiorangelinedfw.pdf

[2] Conrad, Bill (13 July 2011). "DART releases paid parking proposal for Plano station" (http://www.lewisvilleleader.com/articles/2011/07/13/plano_star-courier/news/8953.txt). *Plano Star Courier*. . Retrieved 18 July 2011.

External links

- DART - Parker Road Station (http://www.dart.org/riding/stations/parkerroadstation.asp)

Glenn_Heights_Park_&_Ride

Glenn Heights Park & Ride	
Station statistics	
Address	1200 E. Bear Creek Rd., Glenn Heights, Texas
Connections	DART Route 206 [1] (M-F) DART On-Call Glenn Heights [2] (M-F)
Parking	489 free spaces (No overnight parking)
Other information	
Opened	1990s
Accessible	♿
Owned by	Dallas Area Rapid Transit

Glenn Heights Park & Ride is a small bus-only station located west of I-35E in Glenn Heights, Texas (USA) and is under ownership of Dallas Area Rapid Transit. It is the southernmost transit center in the DART system.

Unlike many Transit Centers, Glenn Heights Park & Ride does not have a heating/cooling facility nor does it have a station attendant, however it does have four passenger shelters.

External links

- Dallas Area Rapid Transit [3] - Glenn Heights Park & Ride [4]

Addison_Transit_Center

Addison Transit Center	

Station statistics	
Address	4925 Arapaho Rd., Addison, Texas
Lines	Cotton Belt *(Future Rail Service for 2030)*
Connections	DART Routes 31 [1], 36 [1], 183 [2], 205 [3] (M-F), 333 [4] (M-F), 347 [5] (M-Sat), 350 [1] (M-Sat), 361 [6] (M-Sat), 362 [7] (M-F), 400 [1], 463 [2], 488 [4] (M-Sat), 534 [8], 536 [9] (M-F)
Parking	300 free spaces (No overnight parking)
Other information	
Opened	1990s
Accessible	♿
Owned by	Dallas Area Rapid Transit

Addison Transit Center is a bus-only station located along Quorum Drive and Addison Road in Addison, Texas (USA). The station opened in 1999.[1] This transit center will become a future rail station on the Cotton Belt as part of DART's 2030 Transit System Plan.[2]

The bus routes within the transit center serves Addison Circle Park as well as The Galleria of Dallas and Valley View Center.

References

[1] Hartzel, Tony (1999-05-25). "DART boosts service in Addison, elsewhere". *The Dallas Morning News*.

[2] "2030 Transit System Plan" (http://www.dart.org/about/expansion/2030plan.asp). DART. . Retrieved 2006-10-24.

External links

- Dallas Area Rapid Transit (http://www.dart.org) - Addison Transit Center (http://www.dart.org/riding/stations/addisontransitcenter.asp)

MLK,_Jr._(DART_station)

MLK, Jr. DART light rail station	
Station statistics	
Address	1412 South Trunk Avenue Dallas, Texas 75210
Coordinates	32°46′26″N 96°45′52″W
Lines	Green Line Orange Line
Connections	DART Routes 12 [1], 26 [2], 50 [3], 202 [4] (M-F) E, and 409 [5] E = Early Morning Trips only
Structure	At grade
Platforms	2 Side platforms
Parking	200 free spaces (No overnight parking)
Other information	
Opened	September 14, 2009[1]
Accessible	♿
Owned by	Dallas Area Rapid Transit
Services	

Preceding station	Dallas Area Rapid Transit	Following station
Fair Park *toward North Carrollton/Frankford*	Green Line	Hatcher *toward Buckner*
	Effective Late 2011	
Fair Park *toward Irving Convention Center*	Orange Line Lawnview branch	Hatcher *toward Lawnview*

MLK, Jr. Station and the adjacent **J.B. Jackson, Jr. Transit Center** is a DART light rail and bus station located on North of Martin Luther King, Jr. Boulevard, between Trunk Avenue and J.B. Jackson, Jr. Boulevard in south Dallas, Texas (USA). It became a Green Line light rail station on September 14, 2009.[2] [3] The station served as the southern terminus of the Green Line until extended to Buckner Station in December 2010.[4]

The station is rich with art and architecture expressing the history of the South Dallas neighborhood and the African heritage claimed by most of its residents.

References

[1] Mosier, Jeff (September 13, 2009). "Dallas rail passengers get sneak preview of new Green Line". *The Dallas Morning News*.

[2] DART Green Line Expansion Information (http://www.ridedart.org/about/expansion/greenline.asp).

[3] Visser, Nancy; Bruce Tomaso (June 9, 2009). "DART hosts riders as it tests span of Green Line project" (http://www.dallasnews.com/sharedcontent/dws/news/localnews/stories/0609dnmetdart.5eeb7bea.html). *Dallas Morning News*. . Retrieved June 9, 2009.

[4] Lindenberger, Michael A. (December 5, 2010). "New Green Line 'starts to complete' DART vision" (http://www.dallasnews.com/sharedcontent/dws/dn/latestnews/stories/120510dnmetgreenmain.1bf8a3c.html). *The Dallas Morning News*. . Retrieved December 7, 2010.

External links

- Dallas Area Rapid Transit (http://www.dart.org) - J.B. Jackson, Jr. Transit Center (http://www.dart.org/riding/stations/jbjacksonjrtransitcenter.asp)

Jack_Hatchell_Transit_Center

Jack Hatchell Transit Center	
Station statistics	
Address	4040 West 15th St., Plano, Texas
Connections	DART Routes 210 [1], 350 [1] (M-Sat), 451 [2], 452 [3], and 566 [2] (M-F) DART Flex - Telecom Corridor (841) [3]
Parking	822 free spaces (No overnight parking)
Bicycle facilities	3 Bike Lockers, 3 Bike Rack
Other information	
Opened	1989
Accessible	♿
Owned by	Dallas Area Rapid Transit

Jack Hatchell Transit Center is a small bus-only station located on 15th Creek Dr. west of Coit in Plano, Texas (U.S.A.). It is owned and operated by Dallas Area Rapid Transit (DART), whose buses mostly serve Plano-area hospitals such as the Presbyterian Hospital and the Medical Center of Plano. It is one of the few DART transit centers outside the Dallas County area.

In 1989 the bus station was opened as the **West Plano Transit Center**. On 12 October 2009, DART dignitaries and the City of Plano renamed and formally dedicated the station in memory of a respected Collin County Commissioner and transportation leader who died in 2008 after a lengthy battle with cancer.[1]

References

[1] Jack Hatchell Transit Center to be formally dedicated (http://www.dart.org/news/news.asp?ID=879) - Dallas Area Rapid Transit (released 7 October 2009)

External links

- Dallas Area Rapid Transit (http://www.dart.org) - West Plano Transit Center (http://www.dart.org/riding/stations/westplanotransitcenter.asp)

Lake_Ray_Hubbard_Transit_Center

Lake Ray Hubbard Transit Center	
Station statistics	
Address	6151 Duck Creek Dr., Garland, Texas
Coordinates	32°50′34.60″N 96°36′5.16″W
Connections	DART Routes 207 [1] (M-F)ST, 283 [2] (M-F), and 385 [3] ST = Select Trips
Parking	660 free spaces (No overnight parking)
Bicycle facilities	2 Bike Lockers, 1 Bike Rack
Other information	
Opened	1990s
Accessible	♿
Owned by	Dallas Area Rapid Transit

Lake Ray Hubbard Transit Center is a small bus-only station located on Duck Creek Dr. west of Broadway (Belt Line Rd.) in Garland, Texas (U.S.A.). It is owned and operated by Dallas Area Rapid Transit, whose buses mostly serve Lake Ray Hubbard as well as South Garland Transit Center and Downtown Garland Station

External links

- Dallas Area Rapid Transit [3] - Lake Ray Hubbard Transit Center [4]

North_Irving_Transit_Center

North Irving Transit Center (Change to North Las Colinas Station pending)	
 Buses awaiting departure at North Irving Transit Center.	
Station statistics	
Address	5601 Riverside Dr., Irving, Texas
Coordinates	32°52′35″N 96°56′02″W
Lines	Orange Line
Connections	DART Routes 202 [4] (M-F), 234 [1] (M-F), 301 [1], 302 [2] (M-Sat), 303 [3] (M-F), 305 [4] (M-F), 306 [5] (M-Sat), 308 [6], 310 [7] (M-F), 311 [8] (M-F), 400 [1] (M-F), 507 [9] (M-Sat), and 528 [3],
Parking	715 free spaces (No overnight parking)
Bicycle facilities	2 Bike Lockers
Other information	
Opened	1990s (Will reopen as Rail Station in 2011)
Accessible	♿
Owned by	Dallas Area Rapid Transit

North Irving Transit Center is a bus-only station located along Northwest Highway (Spur 348) in Irving, Texas (USA). It will be joined with the Irving Convention Center Station when the DART Orange Line light rail line opens in December 2011.[1] There have been rumors that the current transit center will then be sold. [2]

References

[1] DART Orange Line Expansion Information (http://www.dart.org/about/expansion/orangeline.asp).
[2] DART Northwest Corridor LRT Line to Irving/DFW Project (.pdf form) (http://www.dart.org/about/expansion/NWIrvingDFWStationLocationAnalysis17Nov05.pdf)

External links

- Dallas Area Rapid Transit (http://www.dart.org) - North Irving Transit Center (http://www.dart.org/riding/stations/northirvingtransitcenter.asp)

Red_Bird_Transit_Center

Red Bird Transit Center	
Station statistics	
Address	4831 S. Hampton Rd., Dallas, Texas
Connections	DART Routes 21 [1] (M-F), 278 [2] (M-F), and 453 [3]
Parking	598 free spaces (No overnight parking)
Bicycle facilities	1 Bike Locker, 1 Bike Rack
Other information	
Opened	1980s
Accessible	♿
Owned by	Dallas Area Rapid Transit

Red Bird Transit Center is a small bus-only station located on 67 Highway 67 and Hampton Rd. in Dallas, Texas (U.S.A.). It is owned and operated by Dallas Area Rapid Transit, whose buses mostly serve Dallas Executive Airport and Southwest Center Mall within this transit center.

External links

- Dallas Area Rapid Transit [3] - Red Bird Transit Center [4]

Rowlett_Park_&_Ride

Rowlett Park & Ride	
Station statistics	
Address	5000 Martin Drive, Rowlett, Texas
Lines	Blue Line
Connections	DART Routes 207 [1] (M-F), and 283 [2] (M-F) ST DART On-Call East Rowlett [1] (M-F) DART Flex - Garland/Rowlett (887) [2] (M-F) ST = Selected Trips
Parking	326 free spaces (No overnight parking)
Other information	
Opened	1990s (Will reopen as Rail Station in 2012)
Accessible	♿
Owned by	Dallas Area Rapid Transit
Services	

Preceding station	Dallas Area Rapid Transit	Following station
Downtown Garland *toward Ledbetter*	Blue Line	*Terminus*

Rowlett Park & Ride is a bus-only station located on Industrial St. and Martin Dr in Rowlett, Texas (USA). It will become a future DART Blue Line light rail station by 2012.[1]

Unlike many Transit Centers, Rowlett Park & Ride does not have a heating/cooling facility nor does it have a station attendant, however it does have a passenger shelter.

References

[1] DART Blue Line Expansion Information (http://www.ridedart.org/about/expansion/blueline.asp).

External links

- Dallas Area Rapid Transit (http://www.dart.org) - Rowlett Park & Ride (http://www.dart.org/riding/stations/rowlettparkandride.asp)

South_Garland_Transit_Center

South Garland Transit Center	
Station statistics	
Address	4051 Leon Rd., Garland, Texas
Coordinates	32°52′4.99″N 96°39′47.27″W
Connections	DART Routes 164 [1], 372 [2] (M-Sat), 374 [3] (M-Sat), 377 [4] (M-Sat), 380 [5] (M-F), 385 [3], 410 [2], 428 [2], 467 [6], 475 [7], and 519 [8]
Parking	600 free spaces (No overnight parking)
Bicycle facilities	1 Bike Locker
Other information	
Opened	1980s
Accessible	♿
Owned by	Dallas Area Rapid Transit

South Garland Transit Center is a bus-only station located on the I-635 (LBJ Freeway) service road in southern Garland, Texas (U.S.A.). It is owned and operated by Dallas Area Rapid Transit.

External links

- Dallas Area Rapid Transit [3] - South Garland Transit Center [9]

Article Sources and Contributors

LBJ/Central_(DART_station) *Source*: http://en.wikipedia.org/w/index.php?title=LBJ/Central_%28DART_station%29 *Contributors*: A Man In Black, Bevo, Ctgreybeard, D6, DanTD, Dfwcre8tive, Drumguy8800, GETONERD84, Hillrhpc, MattFisher, Mbrstooge, RadicalBender, Saxbryn, Spectre9, Susvolans, The Tom, Wyatt915, 4 anonymous edits

DART_Light_Rail *Source*: http://en.wikipedia.org/w/index.php?title=DART_Light_Rail *Contributors*: DanTD, Delirium, Hailey C. Shannon, JasonLa123, Jllm06, John of Reading, Loodog, Mike1942f, PFMuffinStrike455, Plasma east, Roadrunner76, TexasDawg, The Tom, Useddenim, 8 anonymous edits

High_Five_Interchange *Source*: http://en.wikipedia.org/w/index.php?title=High_Five_Interchange *Contributors*: AndreasPraefcke, Dream out loud, Drumguy8800, EMU CPA, Everything counts, Foobaz, Freewayguy, Furley11, GoldDragon, Greybear1701, Haljackey, Holdercal, Imzadi1979, J.delanoy, Jpbowen, Mapsax, Misaf-Keru, Morriswa, NE2, Novasource, Nutmegger, Onore Baka Sama, Patriarca12, Pepper, ProhibitOnions, Quidam65, Rhatsa26X, Secfan, Sswonk, Tabletop, TexasDawg, Toll booth, TwinsMetsFan, Ufwuct, Zzyzx11, 23 anonymous edits

Interstate_635_(Texas) *Source*: http://en.wikipedia.org/w/index.php?title=Interstate_635_%28Texas%29 *Contributors*: 25or6to4, 78.26, Ansem27, Awtribute, Beland, C.Fred, Champsdfw, DandyDan2007, Dough4872, Drumguy8800, EdH, Freakofnurture, Freewayguy, GETONERD84, Genebob, Gfoley4, Greybear1701, Gridlock Joe, Holdercal, Iluziat, Imperial111, Imzadi1979, Jennavecia, Ligulem, Mes tex, NE2, Novasource, O, Perimeter285, Presidentman, Quidam65, Robertb-dc, Rschen7754, Rufus843, SPUI, Scott5114, Spannermanners, SuperMog2002, TwinsMetsFan, U52983, Vegaswikian, WhisperToMe, WillC, Xnatedawgx, 31 anonymous edits

Central_Expressway_(Dallas) *Source*: http://en.wikipedia.org/w/index.php?title=Central_Expressway_%28Dallas%29 *Contributors*: Alvestrand, Ansem27, Anypodetos, Awtribute, Celestianpower, Common Good, DanTD, DandyDan2007, Debresser, Delldot, Dravecky, Drumguy8800, Explicit, Fredddie, G21a0g8zjs2fsea2gs, Holdercal, Kglavin, Khatru2, Lightmouse, Ligulem, Lovelac7, Mainstream Nerd, Maxim, Mes tex, Morriswa, Msjapan1, Muchness, NE2, NathanBeach, Novasource, Perimeter285, Presidentman, Quidam65, RedPoptarts, Rick Tx, Robertb-dc, Rolypolyman, SPUI, Shawnc, ThaddeusB, Unherdable, Xnatedawgx, 34 anonymous edits

Red_Line_(Dallas_Area_Rapid_Transit) *Source*: http://en.wikipedia.org/w/index.php?title=Red_Line_%28Dallas_Area_Rapid_Transit%29 *Contributors*: Amorymeltzer, Anthony Appleyard, DanTD, Darkcore, Dfwcre8tive, Drumguy8800, GETONERD84, Lost on belmont, Lovelac7, MattFisher, Mbrstooge, NThomas, Osomec, Peter Horn, RadicalBender, SJ Morg, SPUI, Tassedethe, Typhlosion, WhisperToMe, 6 anonymous edits

Orange_Line_(Dallas_Area_Rapid_Transit) *Source*: http://en.wikipedia.org/w/index.php?title=Orange_Line_%28Dallas_Area_Rapid_Transit%29 *Contributors*: Anthony Appleyard, Beagel, C628, Cnbrb, Dfwcre8tive, Drumguy8800, Eliza925, Everyking, Lbbzman, Lincolnite, Lost on belmont, Lovelac7, MattFisher, Mbrstooge, Mrand, Peter Horn, Quidam65, Rjanag, Robertb-dc, Spectre9, Themightywizard, Usgnus, WhisperToMe, XxTrillvillexX9, 9 anonymous edits

Forest_Lane_(DART_station) *Source*: http://en.wikipedia.org/w/index.php?title=Forest_Lane_%28DART_station%29 *Contributors*: A Man In Black, Bevo, D6, DanTD, Dark Serge, Dfwcre8tive, Drumguy8800, GETONERD84, Hillrhpc, MattFisher, Mbrstooge, RadicalBender, Slambo, Susvolans, The Tom, 4 anonymous edits

Westmoreland_(DART_station) *Source*: http://en.wikipedia.org/w/index.php?title=Westmoreland_%28DART_station%29 *Contributors*: D6, DanTD, Dark Serge, Deipnosophista, Drumguy8800, GETONERD84, Hillrhpc, MattFisher, Mbrstooge, Patriarca12, RadicalBender, Susvolans, 2 anonymous edits

Bachman_(DART_station) *Source*: http://en.wikipedia.org/w/index.php?title=Bachman_%28DART_station%29 *Contributors*: DanTD, Dfwcre8tive, Duncan, Mbrstooge, Patriarca12, The Tom, ThomasAndrewNimmo, Typhlosion

Irving_Convention_Center_(DART_station) *Source*: http://en.wikipedia.org/w/index.php?title=Irving_Convention_Center_%28DART_station%29 *Contributors*: DanTD, Dfwcre8tive, Eco84, The Tom, ThomasAndrewNimmo, Train2104, 2 anonymous edits

Spring_Valley_(DART_station) *Source*: http://en.wikipedia.org/w/index.php?title=Spring_Valley_%28DART_station%29 *Contributors*: Anticipation of a New Lover's Arrival, The, D6, DanTD, Dfwcre8tive, Drumguy8800, GETONERD84, Hillrhpc, MattFisher, Mbrstooge, RadicalBender, RoySmith, Spectre9, Susvolans, The Tom, Xnatedawgx, 4 anonymous edits

Parker_Road_(DART_station) *Source*: http://en.wikipedia.org/w/index.php?title=Parker_Road_%28DART_station%29 *Contributors*: D6, DanTD, Dark Serge, Dfwcre8tive, Drumguy8800, GETONERD84, Gaberlunzi, Geoking66, Googuse, Hillrhpc, MattFisher, Mbrstooge, Mpgviolist, RadicalBender, Rich Farmbrough, Rwalker, Slambo, Susvolans, The Tom, 6 anonymous edits

Glenn_Heights_Park_&_Ride *Source*: http://en.wikipedia.org/w/index.php?title=Glenn_Heights_Park_%26_Ride *Contributors*: D6, Mbrstooge, Quidam65

Addison_Transit_Center *Source*: http://en.wikipedia.org/w/index.php?title=Addison_Transit_Center *Contributors*: AndreasPraefcke, CommonsDelinker, D6, Dolenath, Dravecky, Drumguy8800, HMishkoff, Hmains, Mbrstooge, 4 anonymous edits

MLK,_Jr._(DART_station) *Source*: http://en.wikipedia.org/w/index.php?title=MLK%2C_Jr._%28DART_station%29 *Contributors*: D6, DanTD, Dfwcre8tive, Dravecky, Drumguy8800, Duncan, Falso316, Gotsnow, Hmains, Koavf, Lovelac7, Mbrstooge, Moe Epsilon, NThomas, Patriarca12, The Tom, 3 anonymous edits

Jack_Hatchell_Transit_Center *Source*: http://en.wikipedia.org/w/index.php?title=Jack_Hatchell_Transit_Center *Contributors*: D6, Kurykh, Mbrstooge, Raymondwinn, Splooshman, TubularWorld, 5 anonymous edits

Lake_Ray_Hubbard_Transit_Center *Source*: http://en.wikipedia.org/w/index.php?title=Lake_Ray_Hubbard_Transit_Center *Contributors*: Ctempleton3, Mbrstooge, Rjwilmsi, TexasAndroid, TubularWorld

North_Irving_Transit_Center *Source*: http://en.wikipedia.org/w/index.php?title=North_Irving_Transit_Center *Contributors*: D6, Dfwcre8tive, Drumguy8800, Hmains, Hugo999, JustAGal, Mbrstooge, NE2, Slambo, Tim1357

Red_Bird_Transit_Center *Source*: http://en.wikipedia.org/w/index.php?title=Red_Bird_Transit_Center *Contributors*: D6, Mbrstooge, TubularWorld

Rowlett_Park_&_Ride *Source*: http://en.wikipedia.org/w/index.php?title=Rowlett_Park_%26_Ride *Contributors*: D6, DanTD, Drumguy8800, Hmains, Mbrstooge, Slambo, Train2104, 3 anonymous edits

South_Garland_Transit_Center *Source*: http://en.wikipedia.org/w/index.php?title=South_Garland_Transit_Center *Contributors*: Ctempleton3, Mbrstooge, TexasAndroid, TubularWorld

Image Sources, Licenses and Contributors

File:LBJ Central DART Station facing north 2009-12-24.JPG *Source*: http://en.wikipedia.org/w/index.php?title=File:LBJ_Central_DART_Station_facing_north_2009-12-24.JPG *License*: unknown *Contributors*: User:Mang9

Image:Wheelchair symbol.svg *Source*: http://en.wikipedia.org/w/index.php?title=File:Wheelchair_symbol.svg *License*: unknown *Contributors*: ALE!, David Levy, Dream out loud, Ktims, Lensovet, NE2, Sarang, Stifle, Wst, 2 anonymous edits

Image:I-635.svg *Source*: http://en.wikipedia.org/w/index.php?title=File:I-635.svg *License*: unknown *Contributors*: Ltljltlj, SPUI, 1 anonymous edits

Image:US 75.svg *Source*: http://en.wikipedia.org/w/index.php?title=File:US_75.svg *License*: unknown *Contributors*: SPUI, Xnatedawgx, 1 anonymous edits

File:Dallas Area Rapid Transit logo.png *Source*: http://en.wikipedia.org/w/index.php?title=File:Dallas_Area_Rapid_Transit_logo.png *License*: unknown *Contributors*: Duncan, Secondarywaltz

File:DART Light Rail car 113.jpg *Source*: http://en.wikipedia.org/w/index.php?title=File:DART_Light_Rail_car_113.jpg *License*: unknown *Contributors*: User:AEMoreira042281

File:DART Downtown Plano Station 2009-11-23.JPG *Source*: http://en.wikipedia.org/w/index.php?title=File:DART_Downtown_Plano_Station_2009-11-23.JPG *License*: unknown *Contributors*: User:Gaberlunzi

Image:Dfwrailmap2.1.png *Source*: http://en.wikipedia.org/w/index.php?title=File:Dfwrailmap2.1.png *License*: unknown *Contributors*: User:Bchan

Image:Hampton Station 1.jpg *Source*: http://en.wikipedia.org/w/index.php?title=File:Hampton_Station_1.jpg *License*: unknown *Contributors*: DanTD, Drumguy8800, Mang9

File:High Five Interchange.jpg *Source*: http://en.wikipedia.org/w/index.php?title=File:High_Five_Interchange.jpg *License*: unknown *Contributors*: AndreasPraefcke, FlickreviewR, Foobaz, Xnatedawgx, 1 anonymous edits

File:US 75.svg *Source*: http://en.wikipedia.org/w/index.php?title=File:US_75.svg *License*: unknown *Contributors*: SPUI, Xnatedawgx, 1 anonymous edits

File:I-635.svg *Source*: http://en.wikipedia.org/w/index.php?title=File:I-635.svg *License*: unknown *Contributors*: Ltljltlj, SPUI, 1 anonymous edits

Image:High Five.jpg *Source*: http://en.wikipedia.org/w/index.php?title=File:High_Five.jpg *License*: unknown *Contributors*: austrini

File:Interstate 635 map (Texas).svg *Source*: http://en.wikipedia.org/w/index.php?title=File:Interstate_635_map_(Texas).svg *License*: unknown *Contributors*: User:25or6to4

File:I-20.svg *Source*: http://en.wikipedia.org/w/index.php?title=File:I-20.svg *License*: unknown *Contributors*: Ltljltlj, SPUI, 2 anonymous edits

File:I-35E.svg *Source*: http://en.wikipedia.org/w/index.php?title=File:I-35E.svg *License*: unknown *Contributors*: Augiasstallputzer, Ltljltlj, SPUI

File:I-30.svg *Source*: http://en.wikipedia.org/w/index.php?title=File:I-30.svg *License*: unknown *Contributors*: Augiasstallputzer, Ltljltlj, SPUI, 1 anonymous edits

File:US 80.svg *Source*: http://en.wikipedia.org/w/index.php?title=File:US_80.svg *License*: unknown *Contributors*: Bidgee, SPUI, Xnatedawgx, 2 anonymous edits

File:Texas 121.svg *Source*: http://en.wikipedia.org/w/index.php?title=File:Texas_121.svg *License*: unknown *Contributors*: SPUI

Image:635-galleria.jpg *Source*: http://en.wikipedia.org/w/index.php?title=File:635-galleria.jpg *License*: unknown *Contributors*: Robertb-dc at en.wikipedia

File:Dallas, Texas 1955 Yellow Book.jpg *Source*: http://en.wikipedia.org/w/index.php?title=File:Dallas,_Texas_1955_Yellow_Book.jpg *License*: unknown *Contributors*: Infrogmation, Jmabel, SPUI, T2

File:Texas 352.svg *Source*: http://en.wikipedia.org/w/index.php?title=File:Texas_352.svg *License*: unknown *Contributors*: Holderca1

File:Texas 78.svg *Source*: http://en.wikipedia.org/w/index.php?title=File:Texas_78.svg *License*: unknown *Contributors*: SPUI

File:Texas 289.svg *Source*: http://en.wikipedia.org/w/index.php?title=File:Texas_289.svg *License*: unknown *Contributors*: SPUI

Image:Dallas North Tollway.png *Source*: http://en.wikipedia.org/w/index.php?title=File:Dallas_North_Tollway.png *License*: unknown *Contributors*: Admrboltz, Dustimagic, Holderca1, NE2, Pderry

File:North Central Expressway.jpg *Source*: http://en.wikipedia.org/w/index.php?title=File:North_Central_Expressway.jpg *License*: unknown *Contributors*: Original uploader was Ansem27 at en.wikipedia. Later version(s) were uploaded by Duncan at en.wikipedia.

Image:US 175.svg *Source*: http://en.wikipedia.org/w/index.php?title=File:US_175.svg *License*: unknown *Contributors*: Common Good, G.dallorto, SPUI

Image:I-45.svg *Source*: http://en.wikipedia.org/w/index.php?title=File:I-45.svg *License*: unknown *Contributors*: Augiasstallputzer, Ltljltlj, Luigi Chiesa, SPUI, 1 anonymous edits

Image:I-345.svg *Source*: http://en.wikipedia.org/w/index.php?title=File:I-345.svg *License*: unknown *Contributors*: Ltljltlj, SPUI

Image:Texas Spur 366.svg *Source*: http://en.wikipedia.org/w/index.php?title=File:Texas_Spur_366.svg *License*: unknown *Contributors*: Holderca1, SPUI

Image:Texas Loop 12.svg *Source*: http://en.wikipedia.org/w/index.php?title=File:Texas_Loop_12.svg *License*: unknown *Contributors*: Holderca1, SPUI

Image:Texas FM 2786.svg *Source*: http://en.wikipedia.org/w/index.php?title=File:Texas_FM_2786.svg *License*: unknown *Contributors*: Holderca1

Image:Texas 121.svg *Source*: http://en.wikipedia.org/w/index.php?title=File:Texas_121.svg *License*: unknown *Contributors*: SPUI

Image:Texas Spur 399.svg *Source*: http://en.wikipedia.org/w/index.php?title=File:Texas_Spur_399.svg *License*: unknown *Contributors*: Holderca1, SPUI

Image:Texas FM 3038.svg *Source*: http://en.wikipedia.org/w/index.php?title=File:Texas_FM_3038.svg *License*: unknown *Contributors*: User:Mr. Matté

Image:Texas Spur 359.svg *Source*: http://en.wikipedia.org/w/index.php?title=File:Texas_Spur_359.svg *License*: unknown *Contributors*: Holderca1, SPUI

Image:US 380.svg *Source*: http://en.wikipedia.org/w/index.php?title=File:US_380.svg *License*: unknown *Contributors*: SPUI

Image:Texas FM 543.svg *Source*: http://en.wikipedia.org/w/index.php?title=File:Texas_FM_543.svg *License*: unknown *Contributors*: Holderca1

Image:Texas FM 455.svg *Source*: http://en.wikipedia.org/w/index.php?title=File:Texas_FM_455.svg *License*: unknown *Contributors*: Holderca1

Image:Texas FM 121.svg *Source*: http://en.wikipedia.org/w/index.php?title=File:Texas_FM_121.svg *License*: unknown *Contributors*: Holderca1

Image:Texas Spur 381.svg *Source*: http://en.wikipedia.org/w/index.php?title=File:Texas_Spur_381.svg *License*: unknown *Contributors*: Holderca1, SPUI

Image:Texas FM 902.svg *Source*: http://en.wikipedia.org/w/index.php?title=File:Texas_FM_902.svg *License*: unknown *Contributors*: Holderca1

Image:Texas 5.svg *Source*: http://en.wikipedia.org/w/index.php?title=File:Texas_5.svg *License*: unknown *Contributors*: Rocket000, SPUI

Image:Texas FM 1417.svg *Source*: http://en.wikipedia.org/w/index.php?title=File:Texas_FM_1417.svg *License*: unknown *Contributors*: Holderca1

Image:Texas 56.svg *Source*: http://en.wikipedia.org/w/index.php?title=File:Texas_56.svg *License*: unknown *Contributors*: SPUI

Image:Texas FM 131.svg *Source*: http://en.wikipedia.org/w/index.php?title=File:Texas_FM_131.svg *License*: unknown *Contributors*: Holderca1

Image:Texas 91.svg *Source*: http://en.wikipedia.org/w/index.php?title=File:Texas_91.svg *License*: unknown *Contributors*: Kanonkas, SPUI

Image:US 82.svg *Source*: http://en.wikipedia.org/w/index.php?title=File:US_82.svg *License*: unknown *Contributors*: Common Good, SPUI, Xnatedawgx

Image:Texas FM 691.svg *Source*: http://en.wikipedia.org/w/index.php?title=File:Texas_FM_691.svg *License*: unknown *Contributors*: User:Mr. Matté

Image:Texas Spur 503.svg *Source*: http://en.wikipedia.org/w/index.php?title=File:Texas_Spur_503.svg *License*: unknown *Contributors*: Holderca1, SPUI

Image:Texas FM 120.svg *Source*: http://en.wikipedia.org/w/index.php?title=File:Texas_FM_120.svg *License*: unknown *Contributors*: Holderca1

Image:Texas FM 84.svg *Source*: http://en.wikipedia.org/w/index.php?title=File:Texas_FM_84.svg *License*: unknown *Contributors*: Holderca1

Image:US 69.svg *Source*: http://en.wikipedia.org/w/index.php?title=File:US_69.svg *License*: unknown *Contributors*: Bidgee, SPUI, Xnatedawgx, 2 anonymous edits

File:Southbound I-45 at Dallas North Central Expressway.jpg *Source*: http://en.wikipedia.org/w/index.php?title=File:Southbound_I-45_at_Dallas_North_Central_Expressway.jpg *License*: unknown *Contributors*: Original uploader was Ansem27 at en.wikipedia

Image:Central Expressway near NorthPark Center, Dallas, TX.JPG *Source*: http://en.wikipedia.org/w/index.php?title=File:Central_Expressway_near_NorthPark_Center,_Dallas,_TX.JPG *License*: unknown *Contributors*: Original uploader was Ansem27 at en.wikipedia

Image:Central Expressway, Dallas, TX.JPG *Source*: http://en.wikipedia.org/w/index.php?title=File:Central_Expressway,_Dallas,_TX.JPG *License*: unknown *Contributors*: Original uploader was Ansem27 at en.wikipedia

Image:Central Expressway at Haskell Ave, Dallas, TX.JPG *Source*: http://en.wikipedia.org/w/index.php?title=File:Central_Expressway_at_Haskell_Ave,_Dallas,_TX.JPG *License*: unknown *Contributors*: Original uploader was Ansem27 at en.wikipedia

file:Mockingbird_Station_3.jpg *Source*: http://en.wikipedia.org/w/index.php?title=File:Mockingbird_Station_3.jpg *License*: unknown *Contributors*: DanTD, Drumguy8800, Mang9

Image:Cedars Station.jpg *Source*: http://en.wikipedia.org/w/index.php?title=File:Cedars_Station.jpg *License*: unknown *Contributors*: Original uploader was Drumguy8800 at en.wikipedia

Image:DART Red Line @ Union Station.jpg *Source*: http://en.wikipedia.org/w/index.php?title=File:DART_Red_Line_@_Union_Station.jpg *License*: unknown *Contributors*: Original uploader was Drumguy8800 at en.wikipedia. Later version(s) were uploaded by DarkEvil at en.wikipedia.

File:DART's West End Station in the West End Historic District of downtown Dallas.jpg *Source*: http://en.wikipedia.org/w/index.php?title=File:DART's_West_End_Station_in_the_West_End_Historic_District_of_downtown_Dallas.jpg *License*: unknown *Contributors*: Original uploader was Drumguy8800 at en.wikipedia

file:Dallas_Area_Rapid_Transit_2009_Orangelinesymbol.gif *Source*: http://en.wikipedia.org/w/index.php?title=File:Dallas_Area_Rapid_Transit_2009_Orangelinesymbol.gif *License*: unknown *Contributors*: Spectre9

File:Dart orange line construction irving station 2009-08-18.jpg *Source*: http://en.wikipedia.org/w/index.php?title=File:Dart_orange_line_construction_irving_station_2009-08-18.jpg *License*: unknown *Contributors*: User:Mang9

File:Forest Lane DART Station.JPG *Source*: http://en.wikipedia.org/w/index.php?title=File:Forest_Lane_DART_Station.JPG *License*: unknown *Contributors*: User:Dark Serge

File:Westmoreland DART Station.JPG *Source*: http://en.wikipedia.org/w/index.php?title=File:Westmoreland_DART_Station.JPG *License*: unknown *Contributors*: User:Dark Serge

File:Bachman Station.JPG *Source*: http://en.wikipedia.org/w/index.php?title=File:Bachman_Station.JPG *License*: unknown *Contributors*: DanTD

File:Spring valley station 2010-01-16.jpg *Source*: http://en.wikipedia.org/w/index.php?title=File:Spring_valley_station_2010-01-16.jpg *License*: unknown *Contributors*: User:Mang9

File:Parker_Road_Station_-_8-27-10.JPG *Source*: http://en.wikipedia.org/w/index.php?title=File:Parker_Road_Station_-_8-27-10.JPG *License*: unknown *Contributors*: User:Dark Serge

Image:I-35E.svg *Source*: http://en.wikipedia.org/w/index.php?title=File:I-35E.svg *License*: unknown *Contributors*: Augiasstallputzer, Ltljltlj, SPUI

File:Addison Transit Center parking at night.jpg *Source*: http://en.wikipedia.org/w/index.php?title=File:Addison_Transit_Center_parking_at_night.jpg *License*: unknown *Contributors*: Photo: Andreas Praefcke

File:DART N Irving TC.jpg *Source*: http://en.wikipedia.org/w/index.php?title=File:DART_N_Irving_TC.jpg *License*: unknown *Contributors*:

Image:US 67.svg *Source*: http://en.wikipedia.org/w/index.php?title=File:US_67.svg *License*: unknown *Contributors*: Bidgee, SPUI, Xnatedawgx, 2 anonymous edits

MIX
Papier aus verantwortungsvollen Quellen
Paper from responsible sources
FSC® C105338

Printed by Books on Demand GmbH, Norderstedt / Germany